my life cycle

Sunflower

Cherry Lake Press

Published in the United States of America by Cherry Lake Publishing Group
Ann Arbor, Michigan
www.cherrylakepublishing.com

Reading Adviser: Marla Conn, MS, Ed., Literacy specialist, Read-Ability, Inc.
Book Designer: Jennifer Wahi
Illustrator: Jeff Bane

Photo Credits: © amenic181/Shutterstock.com, 5; © Bogdan Wankowicz/Shutterstock.com, 7; © ifong/Shutterstock.com, 9; © directorsuwan/Shutterstock.com, 11; © TaTum2003/Shutterstock.com, 13; © Mironmax Studio/Shutterstock.com, 15; © ER_09/Shutterstock.com, 17; © madmonkey0328/Shutterstock.com, 19; © yevgeniy11/Shutterstock.com, 21; © Try_my_best/Shutterstock.com, 23; Cover, 2-3, 8, 18, 22, 24, Jeff Bane

Cherry Lake Press is an imprint of Cherry Lake Publishing Group.

Library of Congress Cataloging-in-Publication Data

Names: Gray, Susan Heinrichs, author. | Bane, Jeff, 1957- illustrator.
Title: Sunflower / Susan H. Gray ; illustrated by Jeff Bane.
Description: Ann Arbor, Michigan : Cherry Lake Publishing, 2021. | Series: My life cycle | Includes index. | Audience: Grades 2-3
Identifiers: LCCN 2020030671 (print) | LCCN 2020030672 (ebook) | ISBN 9781534179998 (hardcover) | ISBN 9781534181700 (paperback) | ISBN 9781534181007 (pdf) | ISBN 9781534182714 (ebook)
Subjects: LCSH: Sunflowers--Life cycles--Juvenile literature.
Classification: LCC QK495.C74 G73 2021 (print) | LCC QK495.C74 (ebook) | DDC 583/.983--dc23
LC record available at https://lccn.loc.gov/2020030671
LC ebook record available at https://lccn.loc.gov/2020030672

Printed in the United States of America

table of contents

About the author: Susan H. Gray has a master's degree in zoology. She loves writing science books, especially about animals. Susan lives in Arkansas with her husband, Michael. They think that a field of sunflowers is an awesome sight.

About the illustrator: Jeff Bane and his two business partners own a studio along the American River in Folsom, California, home of the 1849 Gold Rush. When Jeff's not sketching or illustrating for clients, he's either swimming or kayaking in the river to relax.

I'm a sunflower seed covered with **soil**. I am not ready to grow yet. It's too cold outside! I wait and wait.

At last, the perfect day arrives. My hard shell covering splits open. My first **root** pokes out.

It grows down, pushing into the soil. My root is looking for water and **nutrients**.

Next, my **shoot** grows straight up into the air and sunlight. Now, I’m a **sprout**!

I keep growing! A flower bud appears on me. It gets bigger and bigger.

One day, it opens up into a huge flower head.

My head is amazing. Look closely. It is packed with tiny **florets**. Some are loaded with **pollen**.

Bees come and walk around on me. They spread pollen everywhere.

Pollen **fertilizes** my florets.
It makes them **develop** seeds.

Soon, those seeds drop to the ground. Leaves fall and cover them. The wind blows soil over them. They wait and wait.

glossary & index

glossary

develop (dih-VEL-uhp) to grow

fertilizes (FUR-tuh-lize-iz) adds material that causes something to grow and develop

florets (FLOR-its) tiny flowers clustered together

nutrients (NOO-tree-uhnts) things such as vitamins that are necessary for life

pollen (PAH-luhn) yellow powdery material produced by flowers

root (ROOT) the part of a plant or tree that grows under the ground

shoot (SHOOT) new plant growth coming up from the ground

soil (SOIL) the ground in which plants grow

sprout (SPROUT) new plant growth from a seed

index